For Mrs. Melissa Eaton and all of the English learners at
Mary Lou Cowlishaw School —K. M.

For River and Jasper —C. S. N.

Library of Congress Cataloging-in-Publication Data:
Names: Messner, Kate, author. | Neal, Christopher Silas, illustrator. Title: Over and under
the canyon / by Kate Messner ; with art by Christopher Silas Neal. Description: San Francisco
: Chronicle Books, [2021] | Series: Over and under | Includes bibliographical references. |
Audience: Ages 5-8. | Audience: Grades 2-3. | Summary: "Over the canyon, the sun scalds
the air, and bakes the desert mud to stone. But under the shade of the canyon hides another
world, where bighorn sheep bound from rock to rock on the hillside, roadrunners make their
nests in sturdy cacti, and banded geckos tuck themselves into the shelter of the sand. This book
takes readers on a journey through the wonders concealed in the curves of the canyon, and all
the secret life hidden in its arms"— Provided by publisher. Identifiers: LCCN 2020041357 |
ISBN 9781452169392 (hardcover) Subjects: LCSH: Animals—California—Anza-Borrego
Desert—Juvenile literature. | Natural history—California—Anza-Borrego Desert—Juvenile
literature. | Ecology—California—Anza-Borrego Desert—Juvenile literature. | Canyons—Cali-
fornia—Juvenile literature. | Anza-Borrego Desert (Calif.)—Juvenile literature. | Anza-Borrego
Desert State Park (Calif.)—Juvenile literature. Classification: LCC QL116 .M47 2021 |
DDC 591.754—dc23 LC record available at https://lccn.loc.gov/2020041357

Manufactured in China.

Book design by Riza Cruz.
Art direction by Amelia Mack.
Typeset in Jannon Antiqua.
The illustrations in this book were rendered in mixed media.

10 9 8 7 6 5 4 3 2

Chronicle Books LLC
680 Second Street, San Francisco, California 94107
www.chroniclekids.com

Over and Under the Canyon

by Kate Messner with art by Christopher Silas Neal

chronicle books · san francisco

High above, Swainson's hawks soar,
circling in the morning-blue sky.

Stones *crunch* under my boots, and stink beetles skitter away. Tiny tracks lead to a crack in the rocks where kangaroo rats rest in the shade. Our trail heads into the canyon.

"What's down there?" I ask.

"Down in the canyon?" Mom says.

"Down in the canyon is a whole world of its own,
where animals find shelter from the desert sun.
We'll feel cooler there, too."

Down to the canyon we climb, step by
careful step. On the hillside above, a
bighorn sheep bounds from rock to rock,
showing us how it's done.

We hike between walls that *stretch* to the sun. Over the canyon, a kestrel glides through a sliver of sky.

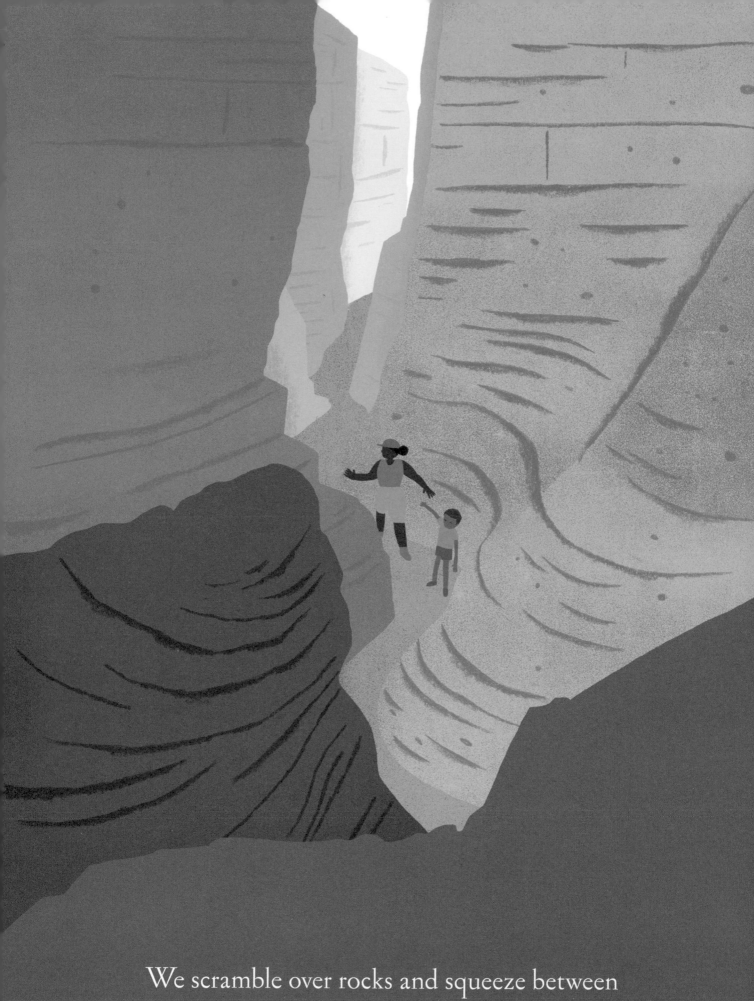

We scramble over rocks and squeeze between boulders. I have to turn sideways to fit!

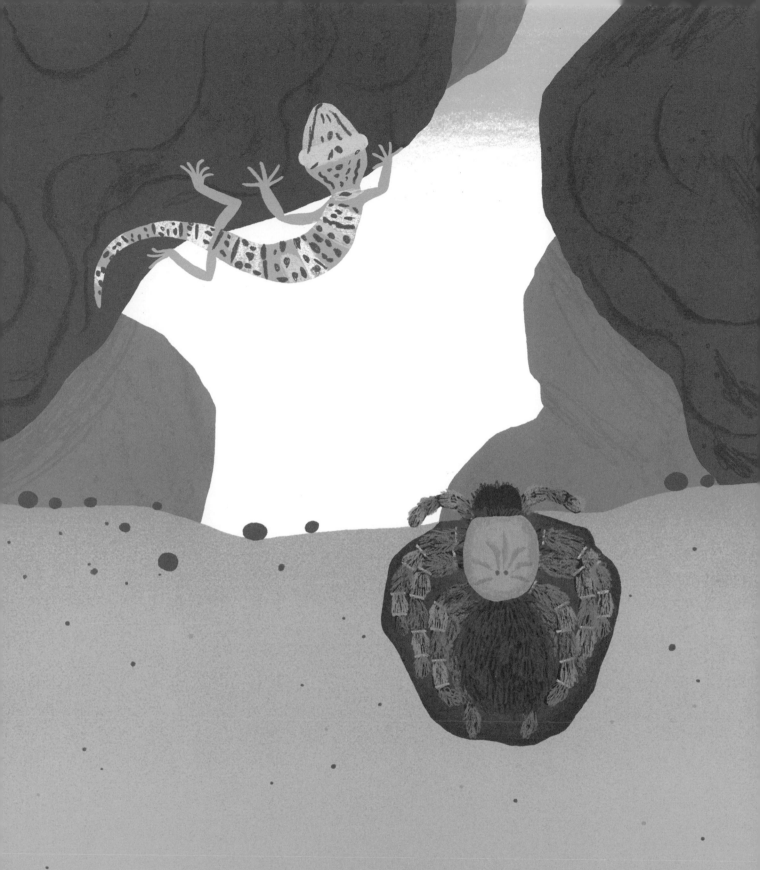

Along the wall, a gecko scoots low to
hide under a ledge. Down in the sand,
a tarantula creeps, folding long, hairy
legs into a crevice of her own.

Over the canyon, a young golden eagle spreads
her wings, hunting for breakfast below.
She *swoops*, and . . .

Whoosh! Brown fur and big ears blur past.

A jackrabbit takes cover in the rocks.
The eagle will go hungry . . . for now.

I'm hungry, too, so it's time for a snack.
When the walls spread out, we climb up
to a cool, shady spot.

"Not there," Mom says.
"That rock's taken."

It is—by a red diamond rattlesnake
whose scales glisten and gleam.

A roadrunner steps close. The snake lifts its tail.

A rattly hum fills the air.

"*Shh . . . ,*" Mom whispers. "*Watch.*"

The roadrunner takes one long-legged
step. Then another. I hold my breath
and wait for the snake to strike. Then . . .

Snap!

The roadrunner catches its head in her beak. She devours her rattlesnake lunch while we munch on raisins and nuts.

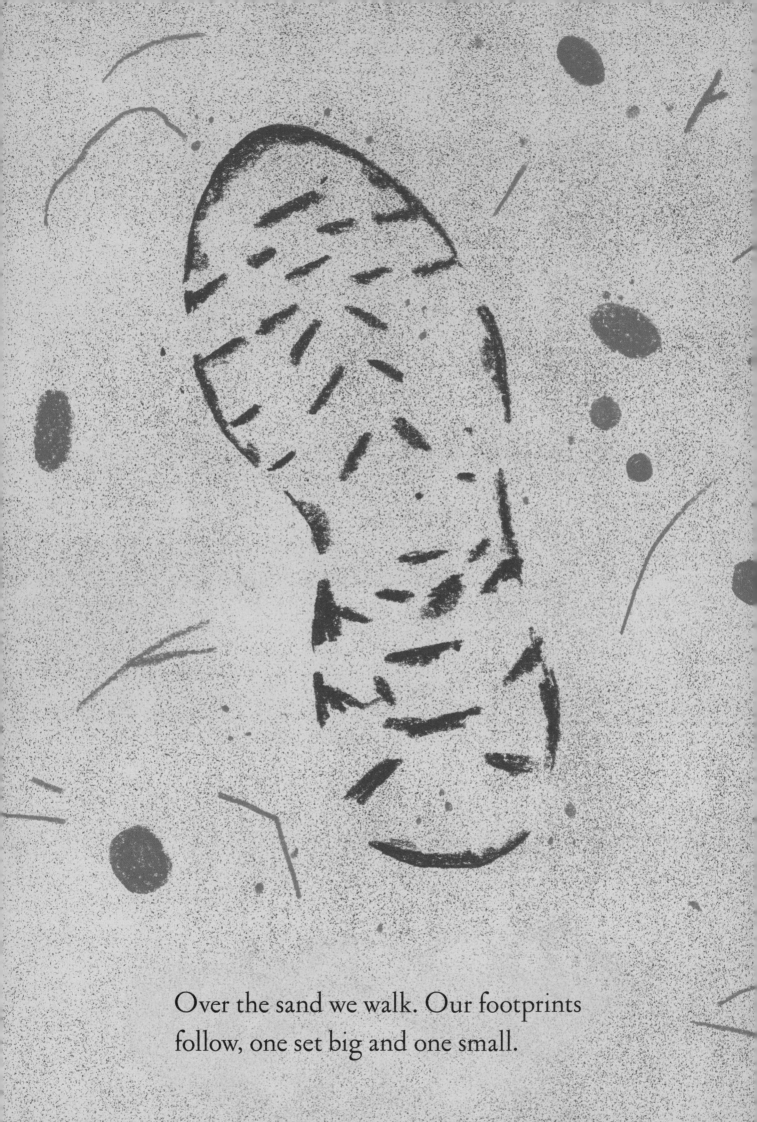

Over the sand we walk. Our footprints
follow, one set big and one small.

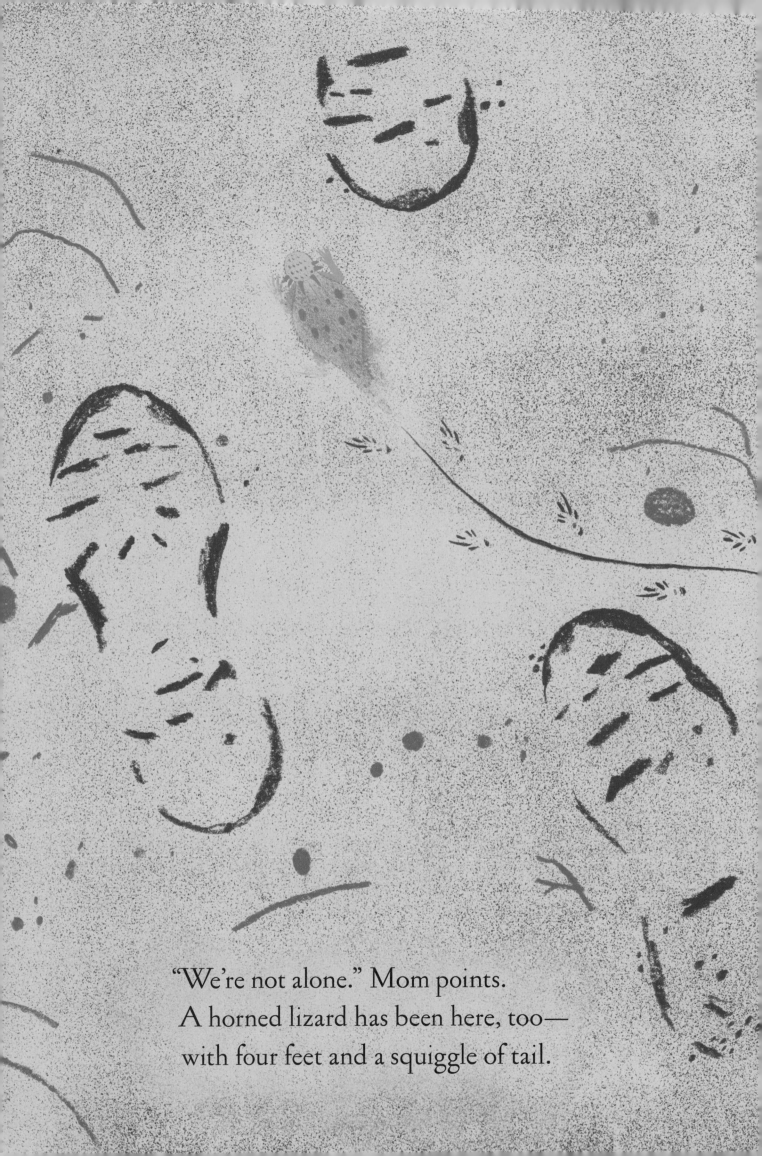

"We're not alone." Mom points.
A horned lizard has been here, too—
with four feet and a squiggle of tail.

On through the sweltering midday we hike, up and down ridges, hot from the sun. A mule deer leaps up and gallops away. We've interrupted his nap.

"When will we be there?" I ask as we
climb a new hill. I'm melting under the sun.

"Right . . . *now.*"

An explosion of wildflowers! Brittlebush,
desert lilies, poppies, and monkey flowers,
blossoming out of the sand.

We wander through the valley, and I dance
between blooms, from yellow to purple to orange.

An antelope squirrel seems to celebrate,
too—darting from bush to bush.

Before we head back, I take a long drink.
I'm ready for rest and some shade.

Down in the sand, harvester ants hide in their home—a cool escape from the sun.

My stomach grumbles as we hike back to camp. I'm not the only one. A desert kit fox wakes to hunt, and cottontails graze in a clearing as sunset paints the sky orange.

Back at camp, Dad's heating up supper,
and our tent looks cozy and cool.

Over the sand, a cactus wren hops through the brush, searching for insects to feed her chicks.

Under the sand, burrowing owls wake to hunt mice.

When the sun fades away, we dig into supper.

Coyotes yip and howl as the twilight turns to night.

"*Watch this,*" Mom whispers.

She points her special flashlight under a
creosote bush, and a hairy scorpion glows
green in the dark.

Bobcats prowl the shadows, and great
horned owls call through the sky.

"Whooo!
Hoo-hoo-hoo.
Whooooo!"

Their day is just beginning as ours comes to an end. I crawl into our tent and close my eyes, and the canyon winds sing me to sleep.

A desert-night lullaby of moonlight and shadows, insect song and stars. And the hidden world nestled away in the canyon.

Author's Note

The seeds for this story were planted way back in 2007, when my family took a trip to the Anza-Borrego Desert State Park in Southern California. We spent a day hiking through washes and slot canyons and observing signs of wildlife. That night, we camped out in the desert to view a total lunar eclipse. As coyotes howled in the distance and the moon slipped into Earth's shadow, the night stars grew even brighter. A desert canyon is truly a magical place to spend a day and night.

About the Animals and Plants

Swainson's hawks are social raptors that are most often found in groups. They eat mammals and small insects, including the caterpillars that feast on desert wildflower blooms. They don't live in the California deserts year-round; they migrate in large groups to South America for the winter and return in spring.

Stink beetles, also called Pinacate beetles, wander the desert floor on warm nights, but by the time the sun comes up, they're usually seeking shelter in burrows made by kangaroo rats or other rodents. Stink beetles eat mostly decaying plant matter. Animals such as ants, roadrunners, and other birds like to eat them, but stink beetles have a secret weapon—special glands that shoot out a terrible-smelling substance! Stink beetles sometimes stand on their heads to warn predators they're about to get sprayed.

Kangaroo rats have strong back legs that make them powerful jumpers. They can leap more than 6 feet (2 metres) in a single bound! Like many desert animals, kangaroo rats tend to stay in their burrows during the hottest part of the day. Their homes under the sand have separate chambers for sleeping, living, and storing food. Kangaroo rats can survive without drinking much water because their bodies get it from the seeds, plants, and insects they eat.

Bighorn sheep are famous for their big curled horns and their climbing skills; they can be seen clambering over desert rocks and ridges. Their split hooves help with balance, and the rough bottoms of their hooves provide a good grip on the rocks. Bighorn sheep eat a variety of desert plants. Males use their horns to fight for mating rights, charging at up to 20 miles

(32 kilometres) per hour. When they crash into one another, it can sometimes be heard from a mile (1.6 kilometres) away!

The **American kestrel** is the smallest falcon in North America, but don't be fooled by its size. It's a fierce predator. Kestrels eat insects as well as small mammals, snakes, and lizards. Because they're small, though, they sometimes end up as prey for larger birds, like hawks and owls. Kestrels that live in the desert tend to nest in holes in the bluffs.

Banded geckos are nocturnal reptiles that spend their desert nights hunting for insects and spiders. They sometimes flick their tongues to pick up chemical clues as they search for prey. Geckos are also prey themselves—for snakes, scorpions, and tarantulas. Like other desert animals, they're in danger of losing their habitat as nearby cities expand.

Tarantulas are large spiders that live all over the world, and a few species make their homes in the deserts of California. They eat insects and sometimes lizards, and hunt mostly at night. Tarantula burrows are lined with silk to keep the sand around them from caving in. Tarantulas are very sensitive to vibrations in the ground, which is why they disappear into their burrows when people approach. They're mostly harmless to people, but if they're bothered, they can shoot prickly hairs out from their bodies that can cause irritation to our skin.

Golden eagles are named for the gold-colored feathers on the back of their heads and necks. Juveniles, like the one in this story, have darker feathers with some white on the wings and tail. Golden eagles are among the largest birds in North America and can have 6- to 7-foot (about 2-metre) wingspans. That's longer than most people are tall! These eagles hunt by soaring over the land, then swooping down to catch prey on the ground.

Black-tailed jackrabbits are easy to identify, thanks to their large ears and long back legs. They can bound up to 20 feet (6 metres) in a single leap and have been known to jump up to 6 feet (2 metres) high! Jackrabbits keep cool in the desert sun by resting in depressions in the sand during the hottest part of the day. Jackrabbits can run up to 30 miles (48 kilometres) per hour, often in a wild zigzag pattern when they're trying to escape from predators like coyotes, bobcats, and hawks.

Red diamond rattlesnakes are venomous and can grow up to 5 feet (1.5 metres) long. They hunt by ambush, waiting for prey to come near before they strike. Rattlesnakes are pit vipers, named for the heat-sensing pits on their faces that help them detect prey. Red diamond rattlesnakes eat lizards, small mammals, and birds. But rattlesnakes can be prey for birds, too. They're known to be a favorite meal for roadrunners.

The **greater roadrunner** can fly when it needs to but prefers to spend most of its time on the ground, where it's known for running at speeds of up to 15 miles (24 kilometres) per hour. Roadrunners nest in sturdy desert bushes or cacti. They aren't picky eaters; their diet includes everything from small mammals to insects and scorpions. Roadrunners are even known to go after rattlesnakes, which they attack by shaking them and pecking them over and over again in the head. Sometimes they team up to do this, so one bird can approach the snake from the front while the other sneaks up from behind.

The **flat-tailed horned lizard** is named for the eight spiky horns that stick up from its head. It looks like a tiny desert dinosaur and can run quickly over the sand. These horned lizards are also experts in camouflage. Their flat bodies and sand-and-stone coloration allow them to blend in with the sand. Horned lizards eat insects and are especially fond of harvester ants, which make up most of their diet.

Mule deer that live in the desert tend to live in the higher elevations and are most active around dawn and dusk. Like most desert animals, they spend the hottest part of the day resting. Mule deer eat a variety of desert plants. Their lips and tongues are specially adapted to let them eat leaves without getting pricked by spines and thorns. Mountain lions are one of the mule deer's main predators.

Wildflowers take over the desert in the spring, carpeting the dry earth with blooms if conditions have been just right. In Anza-Borrego Desert State Park, the right mix of sun, rain, temperature, and wind can lead to an explosion of color. Desert lilies, poppies, monkey flowers, and brittlebush are just a few of the hundreds of species that bloom in the desert.

Antelope squirrels are among the only desert animals that are truly active in the daytime. They're often seen racing from bush to bush. Antelope squirrels eat seeds, fruit from cacti, other plants, and insects. Their many predators in the desert include coyotes, bobcats, snakes, foxes, and birds of prey.

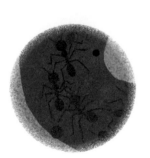

Harvester ants are desert insects that build elaborate underground nests. Their colonies can include more than 10,000 ants that use the nest as a place to stay cool, store food, and lay eggs. From above, a nest often looks like a mound of sand with a hole in the middle, and it's usually surrounded by discarded seed husks.

The **desert kit fox** is the smallest member of the Canidae, or dog, family in North America. These little foxes stay cool by hiding out in their underground dens for most of the day. They come out at night to hunt rodents, rabbits, small birds, reptiles, and insects.

Desert cottontails are common rabbits that often feed in groups at the end of a desert day. They eat grasses and other plants. Like many desert animals, cottontails get much of the water they need from their food. Their predators include coyotes and foxes, as well as bobcats, hawks, owls, and snakes.

Cactus wrens build football-shaped nests in desert plants. They're well adapted to dry conditions and can get by without drinking water at all. Instead, they get all the water they need from the insects and fruit they eat. Grasshoppers are among the cactus wren's favorite foods, and they often pluck off the grasshoppers' wings before feeding the bodies to their chicks.

Burrowing owls are active both day and night. They deal with the heat by roosting in cool burrows. These owls eat mostly insects and rodents and sometimes stockpile food in case they need it later, when they're sitting on eggs and raising chicks. They've been known to store dozens of rodents in their underground nests.

Coyotes are found all over North America and are great at adapting to their environments. They're omnivores, which means they'll eat pretty much anything, including small mammals, birds, snakes, lizards, insects, dead animals, fruits, and other plants. Coyotes can often be heard yipping, yapping, and howling to communicate and keep track of one another.

Anza-Borrego hairy scorpions can grow up to 6 inches (15 centimetres) long. They eat insects, spiders, other scorpions, and sometimes small lizards and snakes. Scorpions are nocturnal, which means they come out mostly at night. They glow a blue-green color under ultraviolet light because of special chemicals in their exoskeletons.

Bobcats are one of the most secretive animals in the desert. They tend to hide by day and hunt by night, approaching their prey slowly and then pouncing with a powerful leap. Bobcats eat mostly cottontails and jackrabbits. However, sometimes they'll also go after a mule deer that's young or weak.

Great horned owls are large birds that can have a wingspan of almost 5 feet (1.5 metres). They're named for the tufts of feathers that stick up from their heads, resembling horns. These birds of prey eat rabbits, rodents, lizards, and scorpions. Their unmistakable hoots are often heard as they call to one another in the desert night: *"Whooooo! Hoo-hoo-hoo. Whooooo!"*

Further Reading

If you'd like to read more about canyon and desert life and ecosystems, you might enjoy the following resources:

Books:

Desert Animal Adaptations by Julie Murphy. Capstone, 2011.

A Desert Scrapbook: Dawn to Dusk in the Sonoran Desert by Virginia Wright-Frierson. Simon & Schuster, 1996.

Grand Canyon by Jason Chin. Roaring Brook, 2017.

Listen to the Desert/Oye Al Desierto by Pat Mora and Francisco X. Mora. Clarion, 2001.

Websites:

Anza-Borrego: Just for Kids, the Desert Natural History Association.
  https://www.abdnha.org/just-for-kids/

Desert Animals & Plants, the San Diego Zoo.
  https://animals.sandiegozoo.org/habitats/desert

Desert Habitat, National Geographic Kids.
  https://kids.nationalgeographic.com/explore/nature/
  habitats/desert/#deserts-camel-sahara.jpg